The Hidden Nature of Power by Xu Dong
Scroll One and Bonus Chapters
Translated by Adam Kryder
Edition 1
© Copyright 2024
ISBN: 979-8-218-50446-5
Little Wing Company | www.littlewingconsulting.co

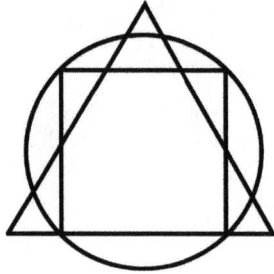

Change Transcends Heaven and Earth

This book is dedicated to Dr. Zhou Shou-Jin

"It would be a great tragedy if you did not pursue the study of this text. Accessing ancient wisdom of this nature will make you aware of things others are not. It may make your life harder for a time, but there is a return..."

虎
鈴
經
。
許
洞
著

何
俊
達
。
譯

The Hidden Nature of Power
Scroll One:
Fundamental Principles
Translation by Adam Kryder

The Hidden Nature of Power
Table of Contents

周守晉博士著 – By Dr. Zhou Shou-Jin
北京大學 – Peking University

The Hidden Nature of Power (Hu Qian Jing, 虎鈐經)[1] is one of the most important military works in Chinese history. It is regarded as an extraordinary book[2], as it takes the principle of Three Phenomena (Heaven, Earth and Mankind, i.e. "三才") into the calculation and utilization of all factors in war; and uses the regulation of Yin(陰) and Yang(陽) to elaborate the original and changing factors involved in warfare. This "extraordinary" nature has had a somewhat limited accessibility to a large part of the society of research for a long time. That is why I was surprised when Adam Kryder came to discuss his translation of this ancient book with me. I decided to engage in the discussion, and immediately after, I saw that what he had done with this formidable work showed fundamental progress in the following aspects.

First of all, he collected three versions of this book (and later found a fourth) to compare with each other. This is a fundamental while rather difficult task. So to think that this book has never been as popular as Sun Tzu or Sun Bin's Art of War, and that the comparison of different versions has rarely been carried out, this makes very little research available for reference. Yet, Adam made comparisons very seriously, and put notes from line to line on his computer and in his notebook. Many of these notes, collations of words and sentences, and even punctuations, turned out to be valuable discoveries.

Secondly, he laid the emphasis of his work on commanding the specific meaning of each word, which gave him much insight into what the author had said, and how it could be conveyed. This can hardly be done by just consulting dictionaries, or without the great efforts he made at the beginning stages.

Last but not least, he himself is well versed in martial arts. His deep understanding of the philosophy and application of martial art played an active role in his communicating with the text of the book and the context as well.

I am extremely pleased to see what Adam Kryder has achieved in this translation of "The Hidden Nature of Power", and I believe his work has made some important breakthroughs in research of the relevant field. Needless to say, our discussions, once or twice a week, have also given me many inspirations.

---

[1] By Xu Dong, 許洞(976-1015), Song Dynasty.
[2] The vice minister of defense in Ming Dynasty, Fan Qin (also well known for his collectives of rare books in his private library Tian Yi Ge) compiled the "Classic of Tiger's Seal" into his series "The Twenty One Most Extraordinary Books (also known as Fan-shi-qi-shu 範氏奇書)".

序文

何俊達著

Foreword
By Adam Kryder

What humanity stands to gain from the application of principle contained in the Ancient Chinese Classics on Military Art is paramount. Yet, when most people think of books on military art, the first work, and usually the only that comes to mind is "The Art of War" by Sun Zi (Sun Tze) (孫子兵法), "The Art of War" is easily the most widespread and well known military work in the world. However, there remains a plethora of little or unknown Ancient Chinese military works that have either been translated and not gained popularity, or have never been translated at all. One of these works is "The Hidden Nature of Power" written by Xu Dong (Hs'u Tong) in the Northern Song Dynasty.

Xu Dong (許洞) (c.e. 976 -1015), was a man from Wu County (吳縣) in Su Zhou (蘇州), his fathers name was Zhong Rong (仲容). Xu Dong was a skilled archer and scholar, and spent three years at Xian Ping (咸平) before becoming a graduate of the imperial examination. He became a heroic warrior and was given a small office, but soon dismissed. At his residence he grew just one stalk of bamboo to represent his one of a kind talent, and inebriated himself daily. He eventually became 'Master of Documents' for Wu Jiang County (烏江縣), and in his daily life entertained the thought that his essays and books were never given proper justice. He was an elite scholar of the "Zuo Shi Zhuan" (左氏傳), and other than "The Hidden Nature of Power" he also authored, a five scroll volume entitled "Profound Explanations of the Spring and Autumn Annals" (春秋釋幽), a ten scroll volume entitled "Expressing Mystery" (演玄), and a "Collection of Literary Works" (文集) which spans one-hundred scrolls in length. All of this was accomplished before his passing at the age of 42. (Song Ren Chuan Ji Zi Liao Suo yin, 宋人傳記資料索引, P.2157).

Thus it is my charge that this ancient scholar's works can no longer be ignored by scholarly research of Ancient Chinese Military Art. The text which I present, (here constituted by the translation of the first scroll), is a synthesis of four different volumes that have been passed down to us through history. The first version used, is "The Hidden Nature of Power" as compiled in the Ming Dynasty (明代) collection of military works entitled, "The Will of Martial Preparations" (武備志) (1619 to 1621). The Second version used, can be found in, "Fan Qin's Collection of the Twenty One Most Extraordinary Works" (範氏二十一種奇書) (1985). The Third version used, is from General Li Yu-Ri's (李浴日將軍) "Special Series of Chinese Military Books" (1957). The fourth version used, was found in the modern compilation entitled "The Perfected Collection of Military Works" (兵書觀止) (1994).

Because each of these books appear at different times in history, and because of the transient nature of most Chinese Classics, there are quite a few differences between the use of characters, punctuation, and included notes. This phenomenon stands as a challenge of providence for any research in the Chinese Classics, and as such the comparisons and contrasts made created a most scrutinous selection of each and every element of the included text. This was done in order to lend the greatest expression of the text's essence, accomplished by a two fold process that utilized the very foundation of the Scholar-Warrior Tradition.

The text continues to be put through the scrutiny of direct experience. With over 20 years of formal training in Chinese and Japanese Internal Arts, I have been able to observe and continue to cultivate the application of many of the martial principles within this work in hand to hand combat training. However, I believe that in the workplace and day to day life is where I have reaped the most benefit. From running restaurants, to international sales management, to executive leadership of organizations, the applications of "The Hidden Nature of Power" have helped me succeed in many complex initiatives. Being so close with this work, I have found myself constantly relying on discovery and curiosity of how these principles relate to the modern world. This is a journey for each reader and practitioner of these ancient methodologies. I want you, the reader, to experience the text, see the connections, and remain ever curious of how this system may adapatavely integrate into your professional and day to day life.

In the beginning, after a year or so of working through my own trial and error with understanding the correct meaning of each and every element of the text, I was privileged enough to meet Dr. Zhou Shou-Jin (Linguistic Anthropology) during my last term at Peking University (周守晉博士。北京大學). The breadth of knowledge, specificity of definition, and generosity of effort imparted upon me (through our dialogues) while evaluating this text, crystallized all of my previous efforts, and brought a life to the text that had been previously unavailable to me. Dr. Zhou and I engaged in vigorous and spirited dialogue, as to the specificity and meaning of each and every element of the text (characters, punctuation, grammar, hidden meanings, etc..; notes of which are made throughout the text and displayed by footnotes) Therefore I can honestly say that the version below is brought to the reader with the greatest care and consideration of the texts original essence, with the foundations of the Scholar-Warrior Tradition fully intact.

Gratitude to Professor Roger DesForges of the State University of New York at Buffalo, and Master Nicholas Logue of the University at Sussex, both of whom helped ensure that my work focus on a previously untranslated text that had solid historical significance.
I know that this text has the possibility to become of even greater benefit in contributing to the further cultural dialogue between the Chinese Classics and the world, and to the betterment of humanity as a whole.

Enjoy!

虎鈴經序　。許洞著　。何俊達譯

The Hidden Nature of Power
Foreword by Xu Dong
Translated by Adam Kryder

虎鈐經者, 將軍之事也。

臣素好奇正之變, 由是而言之也。

然孫子之法奧而精, 使學者難於曉用。

自古兵法多矣。諸家之法膚而淺, 或用者喪於師律。

淺深長短, 迭為表裏。酌中之理, 誠難得焉。

"The Hidden Nature of Power" is meant for Generals, and as it is my intrinsic character is to enjoy the changes between spontaneity and the original state of affairs, thus this is of what I speak. Since Classical times, military methods have been numerous. However, the profound and elite nature of methods of Sun Zi [3] (often) cause those who study them to meet with much difficulty in understanding their use.

The methods of many schools (and experts) are superficial and lacking, and those who use them forfeit military principle. These superficialities and profundities, strong points and shortcomings successively express an intertwined nature. To possess the use of the principles within Sun Zi is indeed a difficult attainment.

---

[3] 孫子 (Sun Zi) or 孫子兵法 (Sun Zi Bing Fa) is commonly known as "The Art of War" much because of its reputation. The same title can also be literally translated as "The Military Methods of Master Sun".

文觀李筌所著太白陰經, 論心術則祕而不言, 談陰陽則散而不備。
以是觀之, 誠非具美。
臣今上採孫子李筌之要, 明演其術, 下撮天時人事之變,
備舉其占, 或作於己見, 或述於古人。

Similarly, to observe all that is contained within Li Quan's[4], "Strategies for the Human Realm" as a mere discussion of a mysterious and unspoken art of mind, and a divulgence on the unrestrained and spontaneous nature of Yin and Yang[5] is indeed an imperfect observation.

Therefore, I will now employ the essence of Sun Zi and Li Quan to clarify and fully express their art. On one hand, extracting the changes of Heaven's Mandate and the Human condition, and on the other hand, comprehensively presenting their fortune. Some examples are made from my own point of view, while others are retold from the classical scholars.

---

[4] 李筌 (Li Quan) author of "太白陰經" Strategies for the Human Realm
[5] See Diagram 1 in the 'Appendix of Diagrams'

名曰虎鈐經，然則奇謀詭道，或不合於六經，既為兵家要用。
故必貫穿條縷以備之。六壬遁甲，星辰日月，風雲氣候，
風角鳥情 雖遠於人事，亦不敢遺漏焉。
至於宣文設奠，醫藥之用，人馬之相，得有補於軍中者，莫不具載？自為一家之
言，創意於辛酉之初。成文於甲辰之末，其書二百一十篇。
分為二十卷。
其年書就於吳郡鳳皇里。
臣洞頓首謹序。

It is called, "The Hidden Nature of Power", and because of its rare strategies and concealed principles, it may not conform to the Six Confucian Classics[6]. Yet since it is of crucial use to military strategists, therefore, it must be well organized and lined up as a self-contained system. Even though astronomical knowledge such as calculating the Six Ren and Hidden Jia[7], the observation of the stars and the sun and moon, the divination of the wind, clouds, inherent energies and climate, the judgment of wind as good or evil, and climate as normal or abnormal have little to do with ordinary human affairs, I also would not dare to abandon or omit them. Even the proclamation and making of announcements, the holding of memorial ceremonies, the utilization of doctors and preparers of medicine, the observation of the appearance of man or horse, all of that which can complement Military Affairs is contained in this book. There is nothing left out.

I created my own system of understanding, initiating its intention at the beginning of the 58th day of the Sexagenarian Cycle[8], and completing its full-text at the end of the 41st day of the Sexagenarian Cycle. This book contains 210 chapters, divided between 20 scrolls. This is the year of this book's completion, in the Phoenix District of Wu County. I, Xu Dong, respectfully bow in earnest account.

---

[6] The Six Confucian Classics: namely <The Book of Changes>, <The Book of Odes>, <The Book of Rites>, <The Book of Music>, <The Book of Histories>, <The Spring and Autumn Annals>
[7] Two prominent Ancient Chinese Divination Methods which date back to the Bone and Shell Scripts (i.e. - 甲古文)
[8]See Glossary

The Hidden Nature of Power
虎钤经

The Hidden Nature of Power
Scroll One

天功第一

Chapter One
Heaven's Force

天道變化，消長萬彙，契地之力，乃有成爾。
天貴而地賤，天動而地靜；
貴者運機，而賤者效力，上有其動而下行其地矣。
是以知天之施地匪專也。
知地之應天有常也。
生機動則應之以生，氣機動則應之以氣，機正則泰，機亂則否。
萬物列形，而否泰交著，見之於地焉，豈止地之為乎？

The Way of Heaven[9] is change, decreasing and increasing the myriad meetings of all phenomena, yet only by uniting with Earth's[10] power may this change be completed. Heaven is the higher and Earth is the lower, Heaven's movement results in Earth's changes. The higher rules destiny and opportunity, while the lower receives the effect of its power. Therefore when Heaven moves, Earth implements its change. From this, it is known that Heaven's offering to Earth is not singular, the conforming of Earth to Heaven has laws. When the opportunity of growth (through Heaven) appears, then (Earth) responds to this growth. When the opportunity of stagnancy appears, then Earth responds with stagnancy. Therefore, righteous[11] driving force leads to peace on Earth, and unrighteous driving force leads to chaos. All arrangements of form, whether chaotic or peaceful, are in a state of constant exchange, and this state is shown on Earth. But, could this (phenomena) be made by Earth alone?

---

[9] Heaven: in "The Art of War" is defined as "Climate and Weather" (Sun Zi, p.4). However it is worthy to note that while Xu Dong's use of this term does conform to its meaning in the "Art of War", he also uses this chapter to explain in detail the extension and further depth.

[10] Earth: in "The Art of War" is defined as "the natural geographic features of earth" (i.e.- distance far or near, dangerous or save passage, etc...) (Sun Zi, p.4) However just as the term "Heaven", Xu Dong's use of this term also extends and brings further depth to the usage of this term.

[11] The use of the word 'righteous' here, need not be perceived as having religious connotation. At its basis it means, 'straight or upright' and may be extended to 'for the benefit' of your troops, mankind, etc..

蓋天道內而地道外者也。

王者, 天也; 將者, 地也。將者, 天也;

士卒者, 地也。

我, 天也;

敵, 地也。

由此觀其所動, 故勝負可知矣。

王之於將也, [12]擇賢受柄, 舉無所疑; 將必內應其正, 外務其順。

應以正則師律嚴, 務以順則臣節貞。

舉而禦敵, 詎有輿尸之患乎？

The Way of Heaven is internal while the Way of Earth is external. The King is as Heaven, the General is as Earth. The General is as Heaven, his Soldiers are as Earth. We are as Heaven, and the opposition is as Earth. By observing the movement of these relationships, defeat or victory can be known. The King's treatment of the General, chooses only the capable to inherit authority. To promote a General without doubt, this General must internally follow righteousness in his action, and outwardly execute his orders with reverence. Following righteousness in his action, yet strictly imposing military discipline, reverently executing, and displaying high personal integrity, loyalty, and devotion. To appoint a General of these qualities to defend against the opposition, how could one still worry about the carriages to transport the dead?

---

[12] (範氏奇書): Fan Qin's Version reads: …, 闔外之寄, … This is omitted because of its lack of necessity to the understanding of this sentence.

君恃智以自用，
倨禮而傲下，
受柄匪人，
使人不信，
將以[13]內抱猶豫之惑，
外喪馭眾之威矣。
舉而禦敵，甯免失律之凶乎？

On the contrary, a King who relies only upon self-willed wisdom, arrogantly treating followers (inferiors), and who give the inheritance of power to an unrighteous man, using him yet not trusting him. This General, will inwardly embrace hesitation and doubt, and will outwardly lose his power over the masses [soldiers]. To appoint a General of these qualities to defend against the enemy, how can one avoid the vicious result of losing discipline?

---

[13] (範氏奇書): Fan Qin's Version reads: "**將不正應**…"while this change might make the meaning more clear, it is unnecessary, the version selected in the text is more closely inline with Ancient Chinese grammar and expression.

師之成敗見之於將焉，豈將之為乎？

將之為任也，智敵萬人，苟無萬人之用，與愚者同矣；

勇冠三軍，苟無三軍之用，與懦者同矣。

善為將者，正而能變，剛而能恤，仁而能斷，勇而能詳，以策馭吏士，未有不振拔勳業，以定禍亂者也。

反是，則吏士外無攻內多離散之勢，勇怯見之吏士焉[14]，豈吏士之為乎？

The defeat or victory of a military's troops can be perceived through its General, yet is this the result of the General's action alone? When the General inherits responsibility, his wisdom must surpass that of multitudes of men; however, if not, he is the same as the ignorant. A General's bravery must surpass that of all his Three Armies[15]; however, if not, he is the same as the cowardly. Those skilled as Generals are timely yet able to adapt, unyielding yet able to empathize, benevolent yet able to judge, brave yet able to be delicate. They use strategy to manage their officers, inspirationally rendering meritorious service, thus naturally suppressing situations of disaster and turmoil. On the contrary, if a General's officials outwardly have no support, they will inwardly divide and scatter. The great force of courage or weakness can be perceived as originating from the officials, but is this just the action of his Officials alone?

---

[14] 武備志 (WuBeiZhi) Reads: "反於是，則吏士外無援內多敵，勢之勇怯，見之吏士焉…"
[15] 三軍 (san jun) Three Armies: The Vanguard, the Main Troop and the Rear Guard

我之於敵也，夫攻拔戰勝，使敵使敢抗衡者，豈敵怯乎？
由我威令整，進退肅，賞罰當也。
覆兵殺將，弱國削地者，豈敵彊乎？
由我不嚴師律故也。
夫如是，亦自上而及下，自內而迨外，其猶天地之用乎。
故天必藉地力也，然後運四氣，正生殺也；
貴必藉賤力，然後能成元功，而建王業也。

As I deal with the opposition, I lay siege to their cities, victorious on all fronts. This is the reason I make the opposition dare not resist. Would this not make the opposition timid and cowardly? Victory is a result of my orderly law and command, strict and impartial advance and retreat, with suitable rewards and punishments. Yet annihilating the troops, killing the General, and crippling the country is the reason for which it will [later] cede. Would I not prefer the opposition to be powerful?

Defeat is caused by my troops lacking strictness and discipline. So, if the structure of power were only dependent from higher to lower, and from inner to outer, then this would be as if Heaven and Earth were without a method use. Therefore, Heaven must draw support from Earth's power, afterwards utilizing the four seasons[16], which impartially judge life and death. The higher must draw support from the lower, only afterwards is the circle completed, and the enterprise of the Kingdom built.

---

[16] 四氣 (Si qi) : a metaphorical use for the four seasons (ie. Spring, Summer, Autumn, Winter)

地利第二

Chapter Two
Earth's Benefit

地之形，險易殊也；地之氣，寒熱異也。

用形與氣，在道逆順焉[17]。

昧此道者，不能得地利必矣。

善用地者，則不然，險而易之，易而險之也，

夫居險用險必內潰，居易用易必外蹶。

當有形之用，逆之者善矣。

Earth's form is differentiated by difficulty and ease of passage. Earth's climate is differentiated by cold and heat. To employ form to deal with climate pertains to knowing its [use of] inversion or obeisance. People ignorant of this principle are unable to obtain Earth's benefit, this is certain!

Those skilled in [this way of] using Earth, think differently [than the ignorant]. If the Earth is difficult in passage, then they [those who are skilled] make it safe. If the Earth is with ease of passage, then they make it dangerous. Therefore, to occupy a position of difficult passage and utilize it as a dangerous position, your unit will most certainly inwardly [i.e. emotionally, energetically, etc.] collapse. To be in a position of ease in passage and utilize it as a safe position, you will most certainly be outwardly defeated [i.e. you will lose the battle, and a great deal of men, etc…]. Facing the use of form, those who invert [its nature] are the skilled.

---

[17] (範氏奇書): Fan Qin's Version reads: "在知逆順焉", omitted due to the literary connotation in the use of the character "道" (Dao)

何謂險而易之？

曰，"簡以制其煩[18]，要以節其動，用以緩其約，不以疏慢為失也。"

What is it called to make an Earth of difficult passage, safe?

It is said, "Use simplicity to control complications, and use necessity as the basis to conduct movement, use utilization of method to loosen restraint. Do not regard comfort and laziness as wrong."

何謂易而險之？

曰，"進止戒嚴，內外無怠，用其謹懼[19]，不以暴急為務也。"

What is it called to make an Earth with ease of passage, dangerous?

It is said, "Advancing and halting with precaution and strictness, inwardly and outwardly without slack, and utilizing restraint and caution and not being aggressive and hasty while pursuing (the goal)"[20]

如是者，乃險易之用也。以方位觀之，則寒熱之氣異也。

Doing as the above, is the utilization of "difficulty and ease" in passage. Use direction and positioning to observe the land, and then the differentiation between cold and hot energies [i.e.- climates etc…] will be clear.

---

[18] (範氏奇書): Fan Qin's Version reads: "簡以夷其政"
[19] (範氏奇書): Fan Qin's Version reads: "慎"
[20] i.e.- minimum damage, and maximum safety to pursue the goal

當有氣之用，順之者善矣。

南方之氣熱，北方之氣寒。

其氣異，則水土之性必相戾，逆諸人而使之飲其地脈，食其土毛，蒙其風氣。瘴癘之疾，凍澀之戾加焉。

Facing the use of inherent energy [21], those who obey [its nature] are the skilled. The southern climate is hot, and the northern climate is cold, their inherent energy is different, so the nature of their water and earth are certain to conflict with each other. As for inverting the nature of people's locale, even while you let them drink the veins of their Earth, eat the vegetation of their soil, and receive the wind's energy; epidemics of malaria, miasma, fever skin ulcers, etc…, these evil[22] diseases will descend upon them.

---

[21] i.e.- climate, weather, seasonal trends, etc…

[22] the use of the word "evil" here pertains to the Chinese medical tradition, where many diseases are caused by what is termed as an "evil wind".

以我之不便，犯順方之人，不有患乎？
天不能以氣順人，君能以人順氣，可也。
是故利不可以專一。
北人之馬，南人之航也，各有便焉。
反是不可措手足焉。
行師者不能擇而用之，斯亦更北南之所便也，冀其成功，遠矣。

With our disadvantages, to intrude upon another people's favorable locale, can there be no sufferings?

Heaven is unable to let climates adapt to Man, but you are able to let Man adapt to a new climates, this is permissible. It is because of this, that benefit is not a singular affair. Northerners are to horses as Southerners are to boats, each possess their own advantage. However, [to place them] in their reverse condition, the people will not know their hands from their feet. Those who lead a traveling military force, and are unable to choose their favorable condition, this is also attempting to change the favorable conditions of all Northerners and Southerners, to still expect success is impossible.

是知地之所利者，可兼而有乎？
善用兵者擇利而從之，善矣。

Therefore, can it be possible to possess the benefit of both?

Those skilled at the use of Military Art, choose benefit and how to follow it, this is skill!

人
用
第
三

Chapter Three
The Utilization of Personnel

今之世取人也，率於[23]其多學，而捨其偏技，非良術也。

兵家所利，隨其短長而用之也。

In present times, to obtain qualified personnel by commanding their only education, and neglecting their eccentric skills, is not a sound method. The beneficial use of Military personnel is to allow their shortcomings and strong points to use themselves.

是以善與者，勿與格鬪，慮其愛人而無勇也[24]。

善保守者，勿與進攻，慮其緩而[25]不猛也。

---

[23] (範氏奇書): Fan Qin's Version reads: "每務"

[24] (範氏奇書): Fan Qin's Version reads: "是以善撫恤者，勿頻鬥，慮其勞疲而無勇也", this version is muddled, and changes the meaning

[25] (範氏奇書): Fan Qin's Version reads:: "…其遲緩而…"

多方者，勿與決事[26]，慮其猶豫也；
多勇者，勿與謀敵，慮其過輕也
精悍者，使鬪。
果敢者，使攻也。
沉毅而性執者，使據阻險；
見小而貪財者，不可使守儲蓄；

It is those who are adept at driving their troops, without having the drive to frequently fight, that in the consideration of mercy are without courage.

It is those adept at protection and defense who do not have the drive for advancement and attack, that in consideration of their defense, are sluggish and un-aggressive.

It is those who are versatile and do not have the drive to decide matters, that in consideration of their versatility, hesitate.

It is those of great courage, who do not have the drive to plan an attack on the opposition, that in consideration of their courage excessively underestimate [the opposition].

Let those who are meticulous and bold, fight.

Let those who are resolute and daring, fight.

Let those who are calm and resolute to occupy strategic points.

Do not let those who are greedy for money and vile of character take care of savings.

智而善斷者，可擇其言。
輕健者，使誘敵；
剛愎者，使當鋒。

---

[26] (範氏奇書): Fan Qin's Version reads: "勿使與於決事" thus complicating the meaning and mistaking the use of the "" thus complicating the meaning and mistaking the use of the "與" character.

利口喋喋者，使其行閒。
善鼠竊狗偷者，使盜號探敵；
惡言多罵者，使之揚毀訾詈；
奇材異識者，使預談論；
崛彊多力者，使斬闢榛莽；
善隨地形結構者，使度樹營柵；

Let the suggestions of those of wise and of good judgment be chosen.

Let those who are light[27] and full of vigor entice the opposition.

Let those who are obstinate defend the front line.

Let those who enjoy long-windedly nagging, practice causation of estrangement between the opposition's own forces.

Let those adept at mouse and dog thievery, investigate and find out the secrets of the opposition.

Let those who often curse with deceitful words, to praise (facetiously), loudly insult, slander and incite severance in the opposition.

Let those of broad vision and generosity comfort the masses.

Let those who are unyielding and of mighty power to cut and remove the strong trees and grasses (to make a road).

Let those skilled at construction according to Earth's form, measure and establish the fence for the barracks.

怯懦者，使輦運器用；
老弱者，使備炊汲；
諳山川，擇高下水泉之利者，使詳地形；

---

[27] i.e.- quick and nimble, etc…

妖言詐辭, 善張皇鬼神之心, 推引天命者, 使揚聲惑眾以動敵心;

善擇地勢平易險巇[28], 知往來細大之蹊路者, 使通糧儲;

奇辭偉辨, 能駕[29]虛矜大者, 使奮振威德;

耳目聰明, 探察敵情者, 使伺候奸偽;

Let those who are fearful and cowardly pull the carriage to transport the weapons for use.

Let those who are old and weak prepare the fires [for food] and fetch water [from the well, etc…]

Let those who have an intimate knowledge of mountains and rivers to choose the benefit of rivers and sources of water, as well as to get detailed information about Earth's form.

Let those of deceitful and lying language, skilled at opening the belief of ghosts and spirits, who calculate and induce Heaven's destiny, to go to intentionally promote and mislead the masses and thereby arouse suspicion in the oppositions heart and mind.

Let those who are skilled at differentiating Earth's terrain as flat and safe, or dangerous and steep, who know the pathways and large roads and all kinds of shortcuts for coming and going, to transport the food stores.

Let those of unusually powerful speech and great debating skills, who are able to manipulate emptiness to create purpose, to bring about the stimulation and upholding of awe-inspiring integrity.

Let those whose ears and eyes are clear and wise, whose duty it is to carefully scout the opponent's situation await [the coming of] spies [of severance and deceit][30].

敏才健筆者, 使主牒檄。

---

[28] (範氏奇書): Fan Qin's Version reads: "阻"
[29] (範氏奇書): Fan Qin's Version reads: "架"
[30] as specified in the above examples of 'spies of severance"

明七曜[31]休咎者，為曆數之士；
善占風雲吉凶者，為候氣之士；
曉六壬[32]遁甲者，為選日時之士；
諳蓍龜者，為卜筮之士。

Let those with wise ability and sturdy pen manage the official documents and declarations.

Let those who understand the fortune and calamity of the seven major "planets" become the officers of calendars.

Let those skilled at forecasting the fortune and calamity of winds and clouds become the officers of climactic energy and the seasons.

Let those who understand astronomical knowledge, such as calculating the Six Ren and Hidden Jia[33]; become the officers of selecting days and time.

Let those who understand the use of grass and tortoiseshell [34]divination, to become the officers of divination and sorcery.

---

[31] 七曜 (Qi Yao):refers to the Five major planets (Mars, Mercury, Jupiter, Venus and Saturn) and the Sun and the Moon
[32] 六壬 (liu ren) : A type of method which utilizes Yin and Yang and the Five elements in the divination of fortune and calamity.
[33] Two prominent Ancient Chinese Divination Methods which date back to the Bone and Shell Scripts (i.e. - 甲古文)
[34] These are used to foresee and tell the coming of fortune or calamity, a divination technique

是四人者，雖推（驗體）測陰陽，各不可使相亂，貴其專一也。
醫藥之人，二十人已上，以兵數增之。獸醫人亦如人醫之數。
大將軍權通材者，與之參議可否。
故我大眾之內，有善有惡，無棄人焉。
人無所棄，斯動則有功矣[35]。

These four types of personnel, although they infer and measure the appearance of Yin and Yang, each is individual and cannot do the work of others, their value lies in their specialties. Doctors and those who prepare the medicine must number twenty-to-one[36]. Veterinary doctors also count in the counting of medical practitioners. Furthermore, Great Generals deeply consider [the opinion of] those who are well versed in medical abilities, when participating in decision making.

Therefore, inside of our great numbers, there are those good and those malicious, but none are useless. When there are no useless personnel, each and every one of their movements will definitely be of good result!

---

[35] (範氏奇書): Fan Qin's Version reads: "斯不怨則動有功矣" thus complicating the meaning.
[36] for every Twenty soldiers there must be One doctor and preparer of medicine.

三才變應第四

Chapter Four
Responding to Change in the Three Phenomena

易曰:見機[37]而作, 不俟終日。

故用兵之術, 知變為大。

軍[38]雖氣銳時勝, (一作遇天時) 而行列散潰, 旌表紊亂, 金鼓不節, 擊之可也[39]。

The Book of Changes states: " When you see the opportunity for action, do not wait for the days finish." Therefore, when utilizing military art, knowing change is a great matter. Even though, Military essence and timing of the opposition is for victory (once written as: "uniting heavens mandate"); when their formations are scattered and chaotic, the large cavalry and the specialized cavalry are disorderly and confused, and the gongs (to withdraw) and the drums (to advance) sound without integrity, you can attack them.

---

[37] 武備志 (Wu Bei Zhi) reads: "幾"

[38]

[39] 武備志 (Wu Bei Zhi) reads: "…不節, 正地肅兵擊之可也。" This difference is omitted due to simplicity and clarity of understanding.

或曰：彼得天時, 詎可破乎

許洞曰：天之所祐, 正也。

怙天時而戾軍政, 與天違也。

天人相違, 不凶何俟？

故兵利以順應順也。

順而逆應之, 必凶之兆也。

Someone asks: "When the opponent gains Heaven's mandate[40], how can you penetrate?"

Master Dong responds: "Heaven's blessing [41]is righteousness! To rely on Heaven's mandate, yet violate Military administration; this is the same as violating Heaven. If Heaven and Man are not in accord, is it not a loss to wait (to attack)? Therefore, martial benefit obeys (to Heaven's mandate, military administration, etc..) the obeisant response. If they use the inverse of an obeisant response, it will certainly foretell vicious defeat.

---

[40] Heaven's Mandate: refers to the benefit of 'Heaven's Force' (see Chapter 1), and/or *right timing* etc…

[41] like most things spiritual, this can be seen with or without religious connotation. A simplified meaning is, *'the right timing'*

或曰：軍乘天時, 跨有地利, 將吏驕怠, 謀畫不精, 軍陣散亂, 如之何？

許洞曰：可擊也。

曰：彼獲天地之利, 如何擊之？

曰：人者天地之心也。苟心不正, 雖有其表, 將焉用乎？

Someone asks: "When the (opponents) Military occupies Heaven's mandate, simultaneously occupying advantageous earth, but the General and Officials are arrogant and lazy, their strategy is without vitality, and military order is relaxed and sloppy, then what can be done?"

Master Dong replies: "Attack!"

"But if the opponent captures Heaven's and Earth's benefit, how can you attack them?"

"People are the heart of Heaven and Earth. If the heart[42] is not righteous, even though having the outer appearance of (Heaven's and Earth's benefit), by what basis will they utilize heart?"

---

[42] an alternate translation for "心" (Xin) is 'mind'

或曰：軍違天時，逆地利。

大將深謀沉毅，部伍清肅，進退有節，如之何？

洞曰：未可擊也。

曰：不獲天地之利，奚謂未可乎？

洞曰：正則可以率天地之用。草寇莽可為王矣。

或曰：彼如是也，我之動將如之何？

許洞曰：先以人，次以地，次以天，然後攻之，必克敵也。

曰：先後如之何？

Someone asks: "When a military force violates Heaven's Mandate and inverts (the nature of) Earth's benefit; yet a Great General deeply plans and is deeply resolute, with his military formations well organized and disciplined, advancing and retreating with integrity, then what can be done?"

Master Dong responds: "One cannot attack."

"Having not captured Heaven's and Earth's benefit, why do you say one cannot strike?"

Master Dong replies, "If righteous, then one can command Heaven's Force and Earth's Benefit for use. Yet, even the armed robbers can become the King's Will."

"When the opponent follows this way of action, what should we do?"

Master Dong replies, "First use Personnel, then use Earth, then use Heaven, attacking after satisfying these (requirements), you will most certainly defeat the opponent."

"With this sequential order, what is to be done?"

曰：利為主。何謂主？

曰：動為客，靜為主。觀敵之動何如，乃應之 。

(所動者，不必戰陣時敵人先動為客也。

但密搆敵人所為之事謂之動。

則我以機應之，□必勝焉。)

先勝而後舉，神明之道也。

"Benefit is the *'Host'*. What is called the *'Host'*?

Master Dong continues, "Action is the *'Guest'*. Tranquility is the *'Host'*. Observe the opponent's action, and then respond. (Those who act during the war, do not think that the enemy's first action is what I called the *'guest'*. Only the affair of secretly constructing the opponent's action is what I call *'action'*. Therefore, when I use the opportunity to respond, I am certain of victory.) First secure victory and then move, this is the spirit that illuminates the Way!"

三才隨用第五

Chapter Five
Following the Use of the Three Phenomenon

天著吉凶，以陰陽辨也；

地布險易，以山川章也；

人包勇怯，以戰陣見也。

苟欲陰陽之順，險易之利，勇怯之用，在於閒可得而擇也。

當彼我相逢，各出不意，忽然交合，曷能擇所利而用哉？

Heaven manifests fortune and calamity, using Yin and Yang to distinguish. Earth declares dangerous and safe passage, using mountains and rivers to make this evident. Mankind possesses bravery and cowardice, using war to make this apparent. If wanting the obeisance of Yin and Yang, the benefit of dangerous and safe passage and the utilization of bravery and cowardice, then only through tranquil implementation can these be made available to suit one's choosing. When the opponent and myself encounter each other, each appearing without previous anticipation, and suddenly engage in combat, how is one able to chose the beneficial utilization of the above circumstances?

或曰：兩師不期逼[43]於險地，天地震晦，雨雪交積，山川不辨。

當此之際，何以禦之？

洞曰：大將止眾堅陣，嚴肅號令。

雖敵來攻，勿與（交）戰。

俟天變少罷。

觀弱彊之勢而後進退之。

Someone asks, "If the two military forces unexpectedly meet and one side approaches a dangerous passage, Heaven and Earth shake in darkness, rain and snow simultaneously accumulate, mountains and rivers cannot be distinguished; facing this opportune moment, how can one defend themselves?"

Master Dong replies, "Great Generals detain their soldiers at a sturdy encampment, asserting command with strictness and integrity. Although the opposition may come to attack, do not engage with them in battle. Wait for Heaven to gradually change and cease in its obstruction. Observe tendencies of weakness and strength[44], and afterward advance or retreat according to these tendencies.

---

[43] Grammatical Ommission: in all versions, there is a character omitted here in order to keep with the structure of the ancient Chinese grammar.

[44] in both the oppositions and your own force.

曰：我既安矣，彼自驚撓[45]，則如之何？

曰：以積[46]兵乘之。

"But, if we are safe, and the opponent themselves are surprised and disturbed, then what can be done?"

"Use your soldiers in wait to engage them."

---

[45] 武備志 (Wu Bei Zhi) reads: "擾" (rao), edited out due to the literary strength and significance of "撓" (nao) character.

[46] 武備志 (Wu Bei Zhi) reads: "精" which would change the translation to "…your best soldiers…"

50

又曰：倉卒之際，大將以何術，即能堅行陣，嚴號令，使士卒[47]師不自驚亂？

"In a sudden moment, what method do Great General's use, to be able to solidify the moving assemblage of his troops, and strictly assert command to make sure his troops do not surprise and scatter themselves?"

---

[47] 武備志 (Wu Bei Zhi) reads: "師" a character change typical of the Ming Dynasty character usage.

洞曰:"善用兵者, 有動必備, 預擇輕勇者二十四人, 八方各三騎, 相去一里, 畫以旗, 夜以鼓, (過山川委曲, 叢林茂密, 畫亦鳴鼓, 恐不見旗[48]。)

第一騎見賊, 畫舉旗。第二騎亦如之。

第三騎馳告, 夜用鼓如畫法。(恐敵聞鼓聲急馳告。)

烟霧翁塞, 風雷震驚, 旗之不見, 鼓之不聞[49], 亟馳告。

斯謂八卦探騎者也。

是以值賊而能備焉。

苟天地有變, 則引輕兵突之。

大陣不可妄動

(凡大兵出[50], 常以輕騎數千人, 別分部隊為游兵, 其將擇驍勇)。

觀彼動靜而後舉焉。故我常用其整也。此皆隨天之變也。"

Master Dong replies, "Those skilled at martial implementation, have certainly prepared for action. Beforehand, choose twenty-four lightly equipped and brave warriors. In each of the eight directions send three on horseback, (each unit) alternately intervaled at 300 paces[51]. In daylight, use the flags[52], and in the night use the drums[53] (If traversing the winding paths of mountains and rivers, and/or thickets and forests which are luxuriant and dense, then strike the drums in the daylight as well, in case the flags are not visible.)

The first of your riders to see the enemy, is to raise the flags in daylight, the second rider also does the same. The third rider then quickly rides to report the situation. At night, utilize the drums according to the daylights method. (This is done in case the enemy hears the sound of the drums, and sends horses to report with greater haste.) In case there is smoke, for, and/or rich tall grasses to obstruct, wind or lightning and thunder to disturb and surprise, thus making the flags unseen and the drums unheard, one must ride with utmost speed to report the situation. This is what is called the

---

[48] After the "旗" character, the 武備志 (Wu Bei Zhi) unnecessarily adds: "之動靜故然爾". This is redundant and misleading.

[49] 武備志(Wu Bei Zhi) shows error as it reads: "旗之不聞"

[50] 武備志reads: "凡兵之行"

[51] 里 (li) is an Chinese measurement constituted by one pace, roughly the equivalent of a modern Meter.

[52] i.e - Establish position

[53] i.e.- advance

method of Ba Gua[54] reconnaissance. This will let one, when meeting with the enemy, simultaneously be able to make preparations.

If Heaven and Earth change, then use your lightly equipped warriors to storm the enemy. The larger military sections cannot be carelessly moved (In the case that the main troops move, normally use a light cavalry of a couple thousand men. Divide the large groups of your soldiers into groups of traveling soldiers, their General will choose the bravest.) Observe the opponents movement, and afterward choose your action. Therefore my normal utilization is ordered. This is all in accordance with Heaven's Changes.

---

[54] 八卦 Ba Gua- The Eight Trigrams, a system of philosophy, divination, for practical application in understanding the natural flow of Yin and Yang.

或曰∶"與敵卒然相遇, 或平原廣澤, 或山谷深峭, 或坡嶝穹隆, 或坳污沮洳, 或
草木蒙密, 俱是危地。當其用之, 如之何？"

洞曰∶"迴渠迂澗, 可以限衝車突馳也。深峽隘口, 可以少擊眾也。長林豐草, 可
以為伏也。原野漫衍, 可以騎兵相屬也。草木隱障, 可以步士接戰。長郊往來, 可
進可退, 可以長戟當敵。踰水跨遠, 高下相乘, 不可得親近, 可以長弩當敵。崖壁
相隱, 狹徑斗迴, 可以劍楯當敵。葦蕭蘆荻, 枝葉朦朧, 可以戈梃, 當敵。此皆隨
地之性也."

Someone asks: "To unexpectedly encounter the oppositions troops on a level, wide-open, low lying field; or a deep and precipitous mountain valley; or a winding and steeply inclined mountain path; or a low lying swampy mountain valley; or on a densely covered wood and grassland; these are all dangerous types of Earth. Facing the use of these different types of Earth, what can be done?"

Master Dong replies, "Tortuous trenches and mountain gullies, can be used to set a limit to the charging carriages and quickly advancing cavalry. A strategic opening at a deep isthmus[55], can be used to use a small number of troops to attack multitudes. Tall trees along with rich and dense grasses can be used to become a point of ambush. A wide-open expansive field can be used as a juncture for cavalries. A hidden and impedient grassy and wooded area can be used for foot soldiers to greet battle. A long broad field in the open air, coming and going where one can retreat or advance can be use for long-handled crescent moon spears[56] to overwhelm the enemy. To cross a wide spanning river, where the height overlaps the laigh, and one is unable to obtain close proximity; one can use (long range) crossbows[57] to overwhelm the enemy. A precipitous cliff where one can hide on a narrow and tortuous pathway, one can use swords and shields to overwhelm the enemy. Where reeds, mugwort, bulrush, phragmites[58]amass, and branches and leaves cause hazed vision, one can use dagger-axes and staffs to withstand the enemy. These situations and their responses all follow the utilization of Earth's nature.

---

[55] a narrow strip of land or water between two mountains
[56] 戟 (ji) long handled weapon (insert drawing)
[57] 弩 (nu) crossbow (more research, drawing, range etc…)
[58] These are all arundinaceous plants

或曰：“寨柵之間，三軍已憩。寇敵乘間衝突攻擊。當此之時如之何？”
洞曰：“使勇者據其前，怯者匿其後。急遽之時，怯者彊驅而前進焉，必挫鋒折銳，不若隱之也。”

"If inside a fenced in military camp, and your three armies[59] recover in rest, yet an invading opposition has infiltrated the camp with an unexpected attack; facing this moment, what can be done?"

Master Dong replies, "If one lets those who are brave occupy the front line, and those who are cowardly support their back, and in a hurried manner the cowardly are forcibly driven to the front line, then this will certainly cause the breaking of the front line and its morale. This method is subservient when compared to hinging together the brave and cowardly."

---

[59] 三軍 (san jun) Three Armies: The Vanguard, the Main Troop and the Rear Guard

曰：當此之時，何暇別勇怯之用乎？

曰：不然。善用兵者，防亂於未亂，備急於未急。結營既定，預擇彊勇者衛外，怯弱者附之，所以慮晦夜之急也。此皆隨人之性也。如是，順天地人預備之道者也。故易之卦以豫[60]者預也[61]。取預備之意為急難之用也。故曰，重門擊柝，以待暴客。蓋取諸豫，用兵者，可不審於此乎？

"When dealing with this situation, is there any spare time to distinguish the use of bravery and cowardice?"

"No, there is not. Those who are skilled at martial implementation, prevent against chaos at times of no chaos, and prepare for haste at times of no haste. When the camp has already been established, choose the strongest and bravest warriors to defend the exterior, and the cowardly and weak to depend on them, this is used in consideration of the dark nights of haste. These uses all follow the nature of Mankind. Therefore, in the Yi Jing's, Ba Gua this is the use of *'Earth below'* and *'Thunder above'* [62]as a foundation. To attain the (original) intention of the foundation in preparation, it will be of utmost importance in situations of haste and difficulty. So it is said, " (Set) several layers of gates and (send bellman to) strike the wooden drum, this is used to wait in preparation for the violent guest."[63]. This meaning must be obtained from the Yu Diagram[64]. For those who utilize military art, how can they not be clear about this?

---

[60] The first of the 64 BaGua Trigrams- its manifestation is "Earth Below and Thunder Above" See Diagram 3 in 'Appendix of Diagrams'

[61] 武備志reads: "故易之豫卦象以豫者預也。"

[62] See Diagram 2

[63] Quote is from《周易》The Book of Changes, See diagram 2

[64] one of the Eight Trigrams of The Ba Gua System, see diagram 2

BONUS CHAPTERS

先謀第十五

Chapter 15
Planning in Advance

用兵之要，先謀為本。
是以欲謀行師，先謀安民。
欲謀攻敵，先謀通粮，欲謀疏陣，先謀地利。
欲謀勝敵，先謀人和。欲謀守據，先謀儲蓄。
欲謀彊兵，先謀正其賞罰。
欲謀取遠，先謀不失其邇。
苟有反，是而用兵者，未有不為損利，而趨害者也。
是故聖王之兵，先務其本。
本壯則末亦從，而茂矣。
苟能知利害之本謀以禦敵，雖有百萬之眾可不勞而克矣。

Using military force requires initial planning as foundation. This is such that when wanting to strategize the mobilization of troops, first plan how to pacify the general public. When wanting to strategize how to attack the enemy, first plan how to transport the provisions. When wanting to strategize how to disperse the encampments, first plan the benefits of earth. When wanting to strategize how to victoriously defend against the enemy, first plan the harmony of men. When wanting to strategize how to safeguard your dependencies, first plan how to stockpile resources and accumulate reserves. When wanting to strategize how to have a strong army, first plan how to right their rewards and punishments. If you want to strategize how to grasp the distant, first plan to not lose that which is near.

If one's strategy runs contrary to this manner of planning their use of military force will forfeit future benefit, and be predisposed to harmful circumstances. It is because of this that the military force of the Sage Kings took primary attainment as their foundation. The state of one's primary foundation is the origin of the end result and diligence. If one is able to understand the foundation of benefit and damage and use it to defend against the enemy, even though they may have one million in manpower, how could you not fatigue and subdue the opponent.

先勝第十六

Chapter 16
Certain Victory

孫子曰：“勝兵先勝，謂先定必勝之術，而後舉也。”
何謂必勝？
許洞曰：“先務三和，次務三有餘，次務三必行。”

Sun Zi says, "Victory in warfare is certain victory. This is called, to first ascertain the method of certain victory, and afterward act upon it."

"What is meant by "certain victory"?

Xu Dong says, "First require the attainment of the Three Harmonies. Next require the attainment of the Three Surpluses. Then require the attainment of the Three Vital Executions."

何謂三和？

曰：“和於國，然後可以出軍。

和於軍，然後可以出陣。

和於陣，然後可以出戰。

不和，則人心離軍。

不和，則教令亂陣。

不和，則行列不正。

先務此三之道，我其可戰耶。”

What is meant by the "Three Harmonies"?

"Bring harmony with the State, and then one can begin enlisting service. Bring harmony with the enlisted service and then one can arrange disposition of combat. Bring harmony with the disposition of combat and then one can go to war. Without harmony, the heart of the people will leave the enlisting of service. Without harmony, the training and command will cause chaos in arranging the disposition of combat. Without harmony, action and arrangement will not be upright. Initially requiring the attainment of the Way of these harmonies, how could I not go to war?"

何謂三有餘？
曰："力有餘，食有餘，義有餘也。力無餘，則困於鬭。
食無餘，則怠於時。
義無餘，則吏士怨。
不務三有餘之術，師其可動耶。"

What is meant by the "Three Surpluses"?

"Ability is in surplus, provisions are in surplus, and sound morals, behavior, and principle are in surplus. If ability is without surplus then one will encounter hardship in battle. If provisions are without surplus then one will be slacken with timing. If sound morals, behavior, and principle are without surplus then the Officials and soldiers will hold resentment. If one does not attain sound method of the Three Surpluses, how can one move troops?"

何謂三必行？
曰："必行其謀，則奸機不成。
必行其賞，則好功者，不愛死。
必行其罰，則有過者，不歸咎。
不必務三必行之道，人其可用耶。"

What is meant by the "Three Vital Executions"?

 "One must execute one's plan, or one's artful capture will not come to be. One must execute one's rewards, or those skilled will not be willing to die. One must execute one's punishments, so those who err will not revert to error"

是以知善務和者，公無松，捨小惠務大惠。
善務有餘者，力諸事而不自怠。
善務必行者，興勇斷，去猶豫之謂也。
舉是九者，務令預定之於前，則萬變千機，然後動乎？
其中矣。
率此以禦，敵未有不勝者也。
故曰，"勝兵先勝者，勝在我也。"
其在夫易曰，"先天不違之義也。"

It is those who are knowledgeable and adept at attaining the Three Harmonies that the elite, without remittance, will aid a small grace in order to obtain a larger one. Those who are adept at attaining the Three Surpluses that power their own affairs, and do not grow weary. And, those who are adept at attaining the Three Vital Executions that awaken brave resolve, and dispel discussion of hesitation.

Those who utilize these nine principles, obtain command by being certain in advance what is to come. Though there are myriads of changes and thousands of skills, after all of this what must they move? Their center! Abiding by these principles to guard against the enemy's attacks, the enemy will incessantly concede victory.

Thus it is said (in Sun Zi), "Those whose victory in warfare is certain victory, are first victorious within themselves." And within the Yi Jing it is said, "Pre-Heaven does not violate sound morals, behavior, and principle."

Appendix of Diagrams

Diagram One:

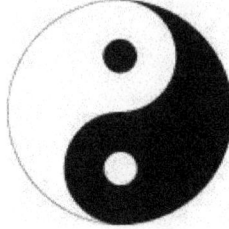

This is a two dimensional representation of the Yin-Yang Symbol (aka-Tai Ji Symbol). In its Three dimensional form it can be found as materially and immaterially functioning at the very basis of all phenomena. Its influence, appearance, effect etc…, appears in the physical, emotional, and spiritual realms.

Diagram Two:

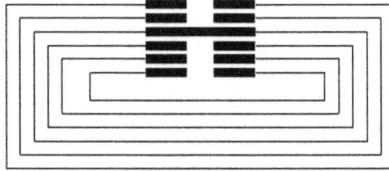

Example of the Layered Gates, and Utilization of the Yu 豫 Diagram mentioned in Chapter 5

Diagram 3

The Eight Trigram Bagua Symbol

Works Cited:

1. Xu Dong. "The Hidden Nature of Power", Ye Shu Ji Cheng Chu Bian, Fan Qin's collection of the Twenty One most Extraordinaryl Books. Zhong Hua Shu Ju Chu Ban Fa Xing, Beijing, China. (qingdai) (1985).

   (In Chinese)
   許洞。"虎鈴經", 業書集成初編, 範氏二十一種奇書 (範氏 奇書)。中華書局發行。北京, 中國。(清代) (1985)

2. Xu, Dong. " The Hidden Nature of Power", Wu Bei Zhi, Di shi liu juan, p. 689-708, Edited by, Fang Feng mao (yuan yi ji).(circa.1594-1630) Taipei, Zong Qing. Hua shi chu ban she, zhong hua min guo ba shi wu nian (1996)

   (in Chinese)

   許洞。"虎鈴經", 武備志, 第十六卷。689 - 708業, 防風茅元議輯。(1594 - 1630年) 台北, 宗青。華世出版社。中華民國八十五年(1996)

   this version is also contained in:

   Zhong guo bing shu ji cheng. wei hui bian . jiefang jun chu ban she (1987)
   中國兵書集成。委會編。解放軍出版社 (1987)

3. Xu, Dong. " The Hidden Nature of Power", Special Series of Chinese Military Books Vol. 6. , Editor: Lee Yu-Ri, Co-editors: Lee Jen-Shiung, Lee-Jen Mao (1957). E-Book by World Military Science Society, 2008 http://www.leeyuri.org/Series.html (In Chinese)

4. Xu, Dong. "The Hidden Nature of Power". Bing Shu Guan Zhi, di er juan. P.686. Fang li zhong zhu bian, Beijing guangbo xueyuan chu ban she 1994
   許洞。"虎鈴經"。兵書觀止, 第二卷, p.686. 房立中主編。北京廣播學院出版社。(1994)

5. Song Ren Chuan Ji Zi Liao Suo yin, p.2157. Edited by chang bi-de, cheng yuan-min, wang de-yi, hou jun-de, hua zhuan zhen chu ban she , zhong hua min guo jiu shi nian (2001)
   宋人傳記資料索引, P.2157. 主編者:昌彼得, 程元敏, 王德毅, 候俊德。華傳真出版社。中華民國九十年 (2001)

6. "Sun zi bing fa". Zhong hua jing dian zang shu, zhu yi zhe: Pian yu-qian deng. Bei jing zhong hua shu ju (2006)
   "孫子兵法"。中華經典藏書。主譯者:駢宇騫等。北京。中華書局 (2006)

7. Taiji symbol: http://hi.baidu.com/liyunxp/blog/item/80a160cfbfd6723df8dc61ca.html
8. Ba Gua Symbol: http://blog.soufun.com/6967912/2460818/articledetail.htm

Glossary:

SAN CAI

The term "三才" (Sān Cái) translates to "Three Powers" or "Three Talents" in English and is a fundamental concept in Chinese philosophy. This term represents three essential elements of the universe:

Heaven (天, Tiān): This element symbolizes the sky, celestial bodies, and the cosmos. It is associated with the overarching principles, natural laws, and forces that govern the universe.

Earth (地, Dì): This element refers to the physical world, the planet, and the environment. It encompasses all things material and tangible, representing the foundation and sustenance of life.

Humanity (人, Rén): This element signifies human beings and human society. It highlights the role of humans within the universe and their ability to influence and harmonize with both Heaven and Earth.

In traditional Chinese thought, these three elements are interconnected and interdependent. The harmony among Heaven, Earth, and Humanity is considered essential for balance, order, and prosperity. This triad is a guiding principle in various aspects of Chinese culture, including medicine, feng shui, and moral philosophy.

THE SEXEGENARY CYCLE AND THE COUNTING OF DAYS:

The sexagenary cycle, or the Stems-and-Branches system (Ganzhi system), is a traditional Chinese method for counting time. It combines ten Heavenly Stems and twelve Earthly Branches to create a 60-unit cycle, used to mark years, months, days, and hours. Here's a detailed explanation of how it works, particularly for counting days:

Components of the Sexagenary Cycle
Ten Heavenly Stems (天干, Tiāngān):
甲 (Jiǎ)
乙 (Yǐ)
丙 (Bǐng)
丁 (Dīng)
戊 (Wù)
己 (Jǐ)
庚 (Gēng)
辛 (Xīn)
壬 (Rén)
癸 (Guǐ)

Twelve Earthly Branches (地支, Dìzhī):
子 (Zǐ) - Rat
丑 (Chǒu) - Ox
寅 (Yín) - Tiger
卯 (Mǎo) - Rabbit
辰 (Chén) - Dragon
巳 (Sì) - Snake
午 (Wǔ) - Horse
未 (Wèi) - Goat
申 (Shēn) - Monkey
酉 (Yǒu) - Rooster
戌 (Xū) - Dog
亥 (Hài) - Pig

Forming the 60-Day Cycle

To form the 60-day cycle, each Heavenly Stem is paired sequentially with each Earthly Branch. The cycle starts with 甲子 (Jiǎzǐ) and continues through all combinations of the stems and branches. The pattern is as follows:

甲子 (Jiǎzǐ)
乙丑 (Yǐchǒu)
丙寅 (Bǐngyín)
丁卯 (Dīngmǎo)
戊辰 (Wùchén)
己巳 (Jǐsì)
庚午 (Gēngwǔ)
辛未 (Xīnwèi)
壬申 (Rénshēn)
癸酉 (Guǐyǒu)
甲戌 (Jiǎxū)
乙亥 (Yǐhài)
丙子 (Bǐngzǐ)
丁丑 (Dīngchǒu)
戊寅 (Wùyín)

...

壬戌 (Rénxū)
癸亥 (Guǐhài)

After the 60th combination, the cycle repeats starting from 甲子 (Jiǎzǐ) again.

Counting Days

When used to count days, the sexagenary cycle assigns each day a unique pair of Heavenly Stem and Earthly Branch, repeating every 60 days. For example, if today is 甲子 (Jiǎzǐ), then:

Tomorrow will be 乙丑 (Yǐchǒu)

71

The day after will be 丙寅 (Bǐngyín)
This pattern continues until reaching 癸亥 (Guǐhài) on the 60th day.
On the 61st day, the cycle starts over with 甲子 (Jiǎzǐ).

<u>Practical Application</u>
This system was historically significant in East Asian cultures for various purposes:

Calendars: Traditional Chinese calendars used the sexagenary cycle to mark years, months,
days, and even double hours (one "hour" in the system is equivalent to two modern hours).
Astrology: The cycle is integral to Chinese astrology, determining zodiac signs and auspicious
times.
Historical Records: Ancient records often used the sexagenary cycle to date events.
In modern times, the sexagenary cycle is still used in traditional contexts, such as in the
Chinese lunar calendar and for cultural festivals.

SIX REN
The term "Six Ren" (六壬, Liù Rén) is associated with traditional Chinese metaphysics and
divination. It refers specifically to a complex system of divination that is one of the three most
renowned methods in Chinese divination, along with the I Ching (Yijing) and Qimen Dunjia.

Overview of Six Ren (六壬, Liù Rén)
Six Ren Divination is an ancient Chinese method of forecasting and decision-making. It has
been used for centuries to provide insights into various aspects of life, including personal
matters, health, weather, politics, and military strategy. The system is intricate and requires a
deep understanding of Chinese cosmology, numerology, and symbolism.

Components of Six Ren Divination
Heavenly Stems (天干, Tiāngān): The same ten Heavenly Stems used in the sexagenary cycle:

甲 (Jiǎ)
乙 (Yǐ)
丙 (Bǐng)
丁 (Dīng)
戊 (Wù)
己 (Jǐ)
庚 (Gēng)
辛 (Xīn)
壬 (Rén)
癸 (Guǐ)
Earthly Branches (地支, Dìzhī): The twelve Earthly Branches, also used in the sexagenary
cycle:

子 (Zǐ) - Rat

丑 (Chǒu) - Ox
寅 (Yín) - Tiger
卯 (Mǎo) - Rabbit
辰 (Chén) - Dragon
巳 (Sì) - Snake
午 (Wǔ) - Horse
未 (Wèi) - Goat
申 (Shēn) - Monkey
酉 (Yǒu) - Rooster
戌 (Xū) - Dog
亥 (Hài) - Pig

Six Lines (六爻, Liù Yáo): This refers to the six lines in a hexagram used in I Ching divination. Six Ren also uses six lines, which can change and interact based on the reading.

Four Pillars (四柱, Sì Zhù): This involves using the year, month, day, and hour of the inquiry to create a chart for divination.

How Six Ren Divination Works
Six Ren divination involves creating a complex chart using the above components. Here's a simplified explanation of the process:

Determine the Four Pillars: The year, month, day, and hour are converted into their respective Heavenly Stems and Earthly Branches.

Establish the Six Lines: These lines are determined by various methods, including casting coins or using yarrow stalks. Each line corresponds to one of the Earthly Branches.

Analyze the Chart: The interaction between the Heavenly Stems, Earthly Branches, and the Six Lines is analyzed. Practitioners look at how these elements interact based on traditional rules and principles to predict outcomes and provide guidance.

Interpret the Results: The final interpretation considers all elements, including the positions and relationships of the Stems and Branches, the changing lines, and the broader context of the inquiry.

Applications of Six Ren Divination
Six Ren divination is used for:

Personal Guidance: Providing insights into personal issues, relationships, and health.
Weather Prediction: Forecasting weather patterns, which was particularly important in agrarian societies.
Military Strategy: Used historically by generals to make strategic decisions.
Political Forecasting: Advising on governance and political matters.
Conclusion

Six Ren (六壬, Liù Rén) is a sophisticated and nuanced method of Chinese divination. It requires a deep understanding of Chinese metaphysical principles and a skilled practitioner to accurately interpret the complex interactions within the divination chart. Despite its complexity, it has been a valuable tool in Chinese culture for centuries, providing guidance and insights across various aspects of life.

HIDDEN JIA

In traditional Chinese metaphysics and astrology, the concept of "Hidden Jia" (藏干, Cánggān) pertains to the hidden or concealed Heavenly Stems within the Earthly Branches. "Jia" (甲, Jiǎ) is one of the Ten Heavenly Stems, and when we talk about "Hidden Jia," we are referring to how the stem Jia can be concealed within certain Earthly Branches.

Earthly Branches and Their Hidden Stems

Each of the twelve Earthly Branches contains one or more of the Heavenly Stems hidden within them. Here's a breakdown of which Heavenly Stems are hidden within each Earthly Branch:

子 (Zǐ) - Rat

癸 (Guǐ)
丑 (Chǒu) - Ox

己 (Jǐ)
癸 (Guǐ)
辛 (Xīn)
寅 (Yín) - Tiger

甲 (Jiǎ)
丙 (Bǐng)
戊 (Wù)
卯 (Mǎo) - Rabbit

乙 (Yǐ)
辰 (Chén) - Dragon

戊 (Wù)
乙 (Yǐ)
癸 (Guǐ)
巳 (Sì) - Snake

丙 (Bǐng)
庚 (Gēng)
戊 (Wù)

午 (Wǔ) - Horse

丁 (Dīng)
己 (Jǐ)
未 (Wèi) - Goat

己 (Jǐ)
乙 (Yǐ)
丁 (Dīng)
申 (Shēn) - Monkey

庚 (Gēng)
壬 (Rén)
戊 (Wù)
酉 (Yǒu) - Rooster

辛 (Xīn)
戌 (Xū) - Dog

戊 (Wù)
丁 (Dīng)
辛 (Xīn)
亥 (Hài) - Pig

壬 (Rén)
甲 (Jiǎ)

Hidden Jia (甲) in Earthly Branches

From the list above, we can see that the Heavenly Stem 甲 (Jiǎ) is hidden within the following Earthly Branches:

寅 (Yín) - Tiger
亥 (Hài) - Pig

Significance of Hidden Jia

The presence of a hidden Jia (甲) in an Earthly Branch has specific implications in Chinese astrology and divination:

Strength and Influence: Jia (甲) represents the energy of Yang Wood (阳木), symbolizing growth, vitality, and initiative. When Jia is hidden within an Earthly Branch, it suggests that these qualities are present but not overtly visible. They can be activated under certain conditions or reveal themselves when needed.

Character and Personality: In BaZi (Four Pillars of Destiny) readings, the hidden stems within a person's Earthly Branches provide deeper insights into their inner character, hidden talents, and latent potential. For instance, if someone has a hidden Jia in their chart, they might possess hidden qualities of leadership, resilience, and creativity that may not be immediately apparent.

Timing and Events: In predictive astrology, the activation of hidden stems, including Jia, can indicate specific times when certain energies or events will surface. This can be influenced by various factors, including transits, progressions, or the interaction with other stems and branches.

Conclusion
Hidden Jia (甲) in the Earthly Branches 寅 (Yín) and 亥 (Hài) represents concealed Yang Wood energy. This concept plays a crucial role in Chinese astrology and metaphysics, offering insights into the underlying qualities and potential events associated with a person or situation. Understanding hidden stems like Jia helps practitioners provide more nuanced and comprehensive readings.

NOTES

NOTES

NOTES

www.ingramcontent.com/pod-product-compliance
Lightning Source LLC
LaVergne TN
LVHW041235080426
835508LV00011B/1218